GADGETS & DEVICES

SCIENCE & TECHNOLOGY

SCIENCE & TECHNOLOGY

GADGETS & DEVICES

Mason Crest

Mason Crest

Mason Crest
450 Parkway Drive, Suite D
Broomall, PA 19008
www.masoncrest.com

Series ISBN: 978-1-4222-4205-6
Hardback ISBN: 978-1-4222-4209-4
EBook ISBN: 978-1-4222-7602-0

First printing
1 3 5 7 9 8 6 4 2

Cover photograph: Dreamstime.com: Olena Korol (phone); Evgeny Baranov (drone); Ocusfocus (VR goggles). Shutterstock: Issarawat Tattong (watch).

Library of Congress Cataloging-in-Publication Data
Names: Mason Crest Publishers, author. Title: Gadgets & devices / by Mason Crest. Other titles: Gadgets and devices
Description: Broomall, PA : Mason Crest, [2019] | Series: Science & technology
Identifiers: LCCN 2018034420| ISBN 9781422242094 (hardback) | ISBN 9781422242056 (series) | ISBN 9781422276020 (ebook)
Subjects: LCSH: Household electronics--Juvenile literature. | Electronic apparatus and appliances--Juvenile literature.
Classification: LCC TK7880 .G335 2019 | DDC 621.381--dc23 LC record available at https://lccn.loc.gov/2018034420

QR Codes disclaimer:

CONTENTS

KEY ICONS TO LOOK FOR

Words to Understand: These words with their easy-to-understand definitions will increase the reader's understanding of the text, while building vocabulary skills.

Sidebars: This boxed material within the main text allows readers to build knowledge, gain insights, explore possibilities, and broaden their perspectives by weaving together additional information to provide realistic and holistic perspectives.

Educational Videos: Readers can view videos by scanning our QR codes, providing them with additional educational content to supplement the text. Examples include news coverage, moments in history, speeches, iconic moments, and much more!

Text-Dependent Questions: These questions send the reader back to the text for more careful attention to the evidence presented here.

Research Projects: Readers are pointed toward areas of further inquiry connected to each chapter. Suggestions are provided for projects that encourage deeper research and analysis.

Series Glossary of Key Terms: This back-of-the-book glossary contains terminology used throughout this series. Words found here increase the reader's ability to read and comprehend higher-level books and articles in this field.

WORDS TO UNDERSTAND

augmented increased or enlarged

aviation having to do with aircraft or flight

commuting traveling back and forth, usually to and from a workplace

connectivity the ability of something to be linked to something

digital expressed as series of the digits 0 and 1

electromagnetic radiation the waves of energy produced by electronic devices, powered by electricity

emergence coming into existence

emit put out, release

estimated calculating anything roughly

film stock specially treated paper-like product that reveals captured photographic images when properly developed, that is, mixed with chemicals and light

floppy disks the first portable computer storage products; thin plastic sheets held in rigid sleeves that could be put into a computer's disk drive to transfer information or programming

gaiter in clothing, a gathered sleeve or leg covering

handwriting recognition the ability of a machine to recognize human writing and turn it into printed or typeset text

immersive completely taking over, completely involving

incredible impossible to believe

licenses accepts a fee under an agreement to allow the use of technology or intellectual property that a company owns

network a group or system of interconnected people or things

pigeon posts letters or other communications sent between two places by being carried by a trained homing pigeon

portable easily movable

prototype a preliminary version that serves as a basis for the later stages

revolutionize to change something radically

rewritable supporting overwriting of previously recorded data

subscriptions the regular payments given for scheduled publications

surveillance the process of watching something for a long time, often from a distance or without the subject's knowledge

telephotography the process of sending a visual image across telephone wires; an early version of a fax process

transaction an agreement in terms of communication or writing between two or more parties

transistor a small device in electronics that has three connections and can link devices or amplify signals

vehicle navigator a device that navigates any vehicle like ship, aircraft, boat, or car

vibrant full of energy and life

wireless transmitting signals without any wire else, usually electronically

INTRODUCTION

The world of technology is moving at a fast pace. Many different types of electronic machines (fun name: "gadgets") have come to rule our lives. These devices perform a specific function for personal or office use. They are the latest and the most futuristic things in our lives. Gadgets may be small or large but each is able to perform a super task.

They have the most recently introduced technologies. These innovative devices are likely to leave a huge impact on our future. Devices such as laptops, **digital** cameras, iPods, printers, scanners, card readers, and PlayStations are all examples of these inventions.

History of Devices

There have been many technological developments in the past few centuries. Newer and better ways of communication and entertainment are being devised every day. The improvements in previous devices gradually paved the way for modern versions. In only a century, we have moved from **pigeon posts**, telegraphs, and letters to mobile phones, computers, and online social **networks**.

Early Telecommunications

In the fifth and sixth centuries, the main modes of distance communications used to be mail and pigeon posts. The process of exchange of information, new ideas, and conversations amongst people was very difficult. In the twentieth century, the telephone, radio, television, computer, and Internet came into existence.

Emergence of Gadgets

These days, modes and means of communication have become convenient. Today there is no need to make a journey by foot or ship to contact people in other parts of the world. All we have to do is turn on our computers or cell phones. Advancements in electronic communication **revolutionized** the way we communicate. Desktop computing is considered the revolution that lead to the **emergence** of modern devices.

SCIENCE FACTS

- Charles Babbage invented the first computer between 1833 and 1877.

- Worldwide, the number of mobile **subscriptions** was **estimated** to be 7.7 billion in 2018.

Action Cameras

People have been using cameras to capture still and moving images of the world around them for more than 150 years. Until the digital age, cameras relied on **film stock** and chemicals to turn the light captured by the cameras into images. Digital cameras were invented in the 1970s, but really came into prominent use in the early 2000s. In 2004, the first GoPro cameras came out and another revolution began. GoPros, and other portable, weatherproof digital cameras, helped people create amazing new films in dramatic places.

Go Anywhere, Film Anything

The key to the GoPro and other action cameras is their durability and portability. Enclosed in weatherproof casing and now even connected wireless to the Internet or smartphones, the cameras can handle water, wind, ice, snow, and more. They are small and can easily attach to helmets and other sports gear. Action sports enthusiasts can now capture the athlete's-eye view of anything they do in the great outdoors.

How to Use GoPro

The action cameras work very much like standard video cameras or videophones. The operator can use buttons to frame and film in still or video mode, in slow-motion or other modes. GoPro is unique in that is can also be attached to gear, such as this surfboard or to atop this windsurfer (below) to capture viewpoints that standard cameras cannot. After the initial "heat" of early GoPros, the company has struggled, but is looking at new models and new filming ideas to continue leading this market.

Laptop

A laptop is a personal computer (PC) designed for versatile and portable use. It inherits all the typical features of a PC, including a display, a desktop, a keyboard, a touch pad, and speakers, all in a single unit. It has a flap which can be folded in order to be carried anywhere. Laptops have a rechargeable battery that charges between three to five hours, and that can be used for the same amount of time in cases of power outage.

Advantages

Unlike a desktop PC, laptops can be used anywhere and anytime. Portability is, of course, the most important advantage. Laptops can be used at home, at an office, and even while **commuting** in cars, trains, and airplanes. They can also be used while sitting in a coffee shop, at lectures, and in libraries. It has easy Internet and local network **connectivity**. It allows an instant access to any information including personal and work files. Laptops are quite convenient to carry, and are power efficient.

Working safely with laptops

Health hazards

Laptops make their users more mobile, but they are blamed for creating health problems. Working on a laptop for hours can cause hazards like neck or spinal pain, strain injury, and skin discoloration. While using laptops, it is recommended to sit up straight with a support to the lower back.

SCIENCE FACTS

- The world's first eye-controlled laptop was launched in March 2011 by Lenovo.

- The Osborne 1 was the first commercially successful portable computer.

Tablet

A tablet is a **wireless**, portable, PC equipped with a touch screen. It is generally smaller than a notebook and larger than a smart phone. There are two basic kinds of tablet personal computers: those with an integrated keyboard (convertible type) and those one without a keyboard (slate type). The main advantage of tablet PCs is their light weight, greater mobility, and easy access to the Internet. Software applications such as web browsers, office suites, and games are built into a tablet. Endowed with advanced features like **handwriting recognition** software, larger memory, better battery life, and wireless Internet access, tablets are considered the best computing option available. Tablets can also be connected to desktop computers as a writing surface.

Write on It

One can easily write on the screen of a tablet using a digital pen and the input utility panel. Writing can be saved in the handwriting of the user or in the typed text format. This amazing technique is possible because of electronic "ink" embedded in the screen.

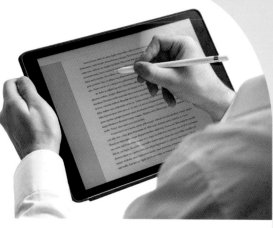

iPad

The iPad, designed and developed by Apple Inc., is one of the latest tablet personal computers available. It does not have a keypad, and is completely controlled by touch display. The first iPad was released in April 2010, and it became a huge success.

SCIENCE FACTS

- The idea of tablet computing was sketched out in 1971 by Alan Curtis Kay, an American computer scientist.

- A rugged-type tablet PC is built with a tough exterior and a shockproof hard drive so that it can be used by construction workers and military personnel.

Scanner

In computing, a scanner is a device that scans pictures, images, texts, handwriting, or an object, and converts it to a digital image. From the 1920s to the mid-1990s, **telephotography** and fax input devices were used to send pictures or text and handwritten messages. There are many types of scanners, such as flatbed scanners (commonly used in offices), drum scanners, film scanners, and handheld scanners.

Uses

A good scanner, unlike the early photocopiers and scanners, helps in a speedy **transaction**. It performs the function of storing important data. It reduces paperwork and saves money. It can share documents right away and scan photos for printing.

Features

Almost all scanners are very easy and convenient for office and personal use. Many can hold up to fifty pages and can scan books and 3-D objects. They have a fast preview screen, and the power to load original copies with one- or two-sided paper, and work at the press of a button. The world's fastest scanner, called the CzurTek scanner, can scan up to five hundred pages in seven minutes, and has up to 2GB of storage space.

- Scanning technologies have now launched 3-D scanners with digital cameras.

- The Dacuda PocketScan is the world's smallest full-page scanner. It is like having a flatbed scanner in the palm of your hand.

Hand-held Scanner

Scanners can read more than documents and photos. The bar codes found on just about every consumer product can be read with a handheld scanner like this one; the laser light translates the lines into information needed by the retailer or the shipper.

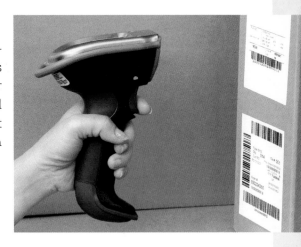

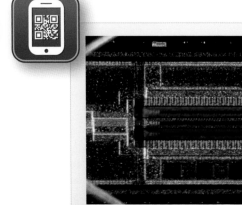

How scanner tech works

Smartphone

Smartphones are wireless tele-communication devices that are portable and can be used anytime and anywhere. Smartphones provide additional services like Short Message Service (SMS), or text messaging, and Multimedia Messaging Service (MMS), or multimedia services. Internet access, Bluetooth, and infrared services are additional services provided in almost every advanced smartphone.

Incredible Gadget

Mobile phones are one of the most **incredible** inventions of all times. These little gadgets have not only made communication easy, but they have also provided us with quick and simple access to the world of the Internet, games, music, and videos. The greatest advantage of a mobile phone is its mobility—it goes wherever you go.

Invention

The first practical mobile phone was invented in 1971 by a Motorola researcher, Martin Cooper. Cooper's phone was called the Motorola DynaTAC, and weighed about 2.5 pounds (1.1 kg). Cooper made the first phone call on April 3, 1973.

Health Hazard

Like all good things, mobile phones come with their share of drawbacks. The most threatening, of course, are its negative impact on health. Studies have shown that **electromagnetic radiation** used by mobile phones can cause severe diseases like cancer. The World Health Organization (WHO) classified mobile phone radiation as possibly carcinogenic (cancer causing). Prolonged exposure to these radiations can damage the brain cells and the eyes.

SCIENCE FACTS

- An industry survey predicts that nearly 3 billion people will have a smartphone by 2020.

- The Apple iPhone 7 Plus has the greatest internal memory in the world with 256GB of storage capacity.

Bluetooth Audio

Smartphones led to the development of a technology that has revolutionized how we listen to music and how we interact with our phones. Bluetooth tech wirelessly connects devices such as headphones, portable speakers, and phones.

Wire-free Tunes

Wireless Bluetooth headphones connect to a person's smartphone. The signal only works within a narrow distance range, so that devices don't overlap. The device is "paired" with the smartphone. Whether listening to music, an audiobook, or a podcast, Bluetooth makes sounds portable and more fun than ever.

Bluetooth tech: How it works

Walk and Talk

Businesspeople love their wireless earbuds. These Bluetooth-enabled devices let them use their phones hands-free. On the go, this can mean not missing time from work while traveling. At the office, it makes working on a computer or writing notes much easier. Headsets and earbuds like these are universal among most businesses.

SCIENCE FACTS

- Jaap Hartsen designed the tech that became the first Bluetooth headsets. The Bluetooth SIG company now owns the brand and **licenses** its use.

- Tech trivia: Bluetooth sends its low-power radio signal on the 2.45 GHz frequency.

SD Cards

Multi-media cards read by Secure Digital (SD) slots on computers and smartphones are postage stamp–sized devices used for storage. They are used in mobile phones, digital cameras, digital audio players, and personal digital assistants (PDAs). They perform functions like copying pictures taken from digital cameras and copying songs and other data downloaded from the Internet.

Storing Facility

The most important aspect of an SD card is its storage capacity. These user-friendly cards can store music files, photographs, videos, and movies. They have replaced CDs and digital video discs, or DVDs. The large storing capacity allows its users to watch movies and listen to music while traveling. The storage capabilities of SDs have made mobile phones young people's best friend.

Types of MMCs

There are many variations of these cards. The Reduced-Size Multi-MediaCard (RS-MMC) is a small MMC that was introduced n 2004. It was used in many mobile phones. The MMCmicro is even smaller than the RS-MMC. The microSD is an even smaller MMC, now used in most smartphones. Other types of MMCs include SecureMMC, Embedded MultiMediaCard (e-MMC), and others.

SCIENCE FACTS

- SanDisk's 1TB microSD card has the highest internal storage capacity in the world.

- A PocketStation is considered a PlayStation Memory Card.

Flash Drive

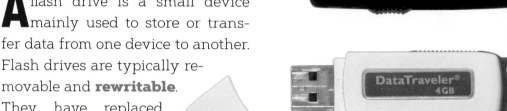

A flash drive is a small device mainly used to store or transfer data from one device to another. Flash drives are typically removable and **rewritable**. They have replaced **floppy disks**, CDs, and DVDs. They are also known as pen drives, USB drives, thumb drives, or pocket drives. Many flash drives have password security. They are more convenient for transferring files between computers at home or office.

Development of USBs

The first USBs became available in the year 2000. They were sold by IBM and Trek technology under the names "DiskOnKey" and "Thumbdrive." These two flash drives paved the way for modern USBs, which come with greater speed for file transfer and reading.

Features

Flash drives are not affected by scratches and dust. They are small and light, available in many shapes, sizes, and colors, and can be attached to a keychain. They are rewritable and require no internal power. Some USBs also have an internal memory slot and allow expandable storage.

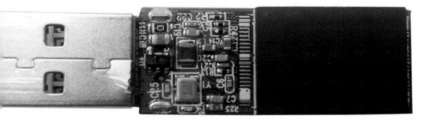

SCIENCE FACTS

- SanDisk's 1TB USB-C flash drive is the smallest pen drive **prototype** in the world, and can store up to 1TB of data.

- The Shawish Geneva Magic Mushroom USB pen drive is the most expensive pen drive.

Smart Watch

A smartwatch is a watch packed with a wide range of features. These watches are best known for their appealing design and comfort. They are designed with supreme functionality and modern technology, and are very easy to understand and operate. There are different types of watches offering features like Bluetooth, camera, Wi-Fi, MP3 player, and an ability to synch with smartphones.

Various Gadget Watches

- **Calculator watch:** The calculator watch was one of the earliest gadget watches. It came equipped with a calculator and was highly popular.
- **Hiking watch:** A watch aptly made for hikers, it comes equipped with an altimeter, barometer, and digital compass.
- **Photo album watch:** These watches come with the ability to display photographs on the background.
- **Watch phone:** These wrist watches double as a smartphone too.

Diving watch

All common watches come with a warning, "Keep away from water." But a diving watch is not damageable by any kind of water pressure.

SCIENCE FACTS

- Timex has launched a unisex gadget watch, which allows continuous tracking of one's heart rate.

- The Polar AW200 technology watch is a watch that measures the quantity and quality of walking exercise by listening to one's body.

Video Game Console

Video game consoles are a popular form of entertainment. They are mobile devices that are designed for single or multiple users, and can be attached to a monitor or a television screen for display. Consoles are specialized computers in themselves and are often based on the same CPU that computers have. The first video game console was designed by Ralph Baer, a German-born television engineer in 1967. Baer's game worked on a standard television. It later came to be known as the Brown Box.

PlayStation

The PlayStation is a video game console developed by Sony. The first generation PlayStation was launched in 1994, followed by the second generation in 2000, and the third generation in 2006. PlayStation **Portable** (PSP) was released in 2005. Sony's latest PlayStation 4 was released in 2014.

- The Magnavox Odyssey was the first commercial video game console. It was developed in 1972.

- The Game Boy introduced in 1989 by Japanese multinational company Nintendo, was the first major handheld console.

Features of PSP

PSP allows its users to access games, videos, music, photos, and the Internet at any time, and to share them with friends. They are sleek, stylish, portable, and perfectly pocket sized. They have a built-in microphone, which allows its users to chat around the world using Skype. It allows gamers to play PlayStation games anywhere in the world, and allows the downloading of games via Wi-Fi. It has dual thumb-stick controls.

Virtual Reality Goggles

The idea of creating a fully **immersive** environment has been around for a long time. In recent years, however, the tech has caught up to the idea. Now consumers can buy virtual reality (VR) goggles. By wearing them, they can interact in a new and stunning way with entertainment.

What You See Is Not Real

People wearing VR goggles get a 360-degree view of the world that is projected inside the screens. When the person moves, the scene "moves," too, following their body movements.

SCIENCE FACTS

- The Oculus Rift VR device was the first to gain commercial success. Many companies have jumped on the bandwagon since.

- Military training organizations use VR to help prepare soldiers for what they might see in combat.

Experience virtual reality . . . sort of.

Game Play Rules

Though VR can be used for training, its most popular—and lucrative—use is in gaming. Users go beyond clicking a mouse to actually moving their whole bodies to interact with the game characters and environment. Additional gear (such as gloves, bodysuits, and even a treadmill) helps create even more touchpoints for the user. The experience can be scary for some users, and others report experiencing motion sickness as their brains learn to deal with the new way of getting information and visual stimuli.

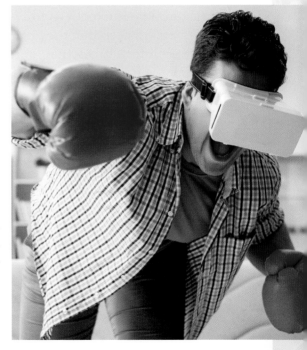

Electronic Clothing

Electronic clothing, or e-textiles (electronic textiles), have built-in digital components such as LEDs, wires, and even **transistors**, diodes, and solar cells. The information that the clothing gathers can be taken in by the user to help with fitness or other needs.

Electronic Jackets

The first electronic jacket was released in the first half of 2007. These jackets have modern technology devices built into them. They have the capability of protecting its user under all sorts of weather conditions. They can deliver iPod music through Bluetooth technology, and are designed so that they can deliver more comfort, protection, and entertainment to its users.

Features

These jackets are waterproof and have a breathable shell. They have a voice dial button for making phone calls. They have a waist **gaiter** and an adjustable cuff closure. They also have goggle pockets and oversized organizer chest pockets. They feature stereo speakers and a microphone, with the speakers built into the hood and the microphone built into the upper section near the collar. They are available in all sizes for both men and women.

FitBit

When the FitBit first came out, it was a big hit. People saw the fitness-activity-monitoring device as a way to encourage an active lifestyle. The device calculated heart rate, calorie usage, movement repetitions, and more. The fad has faded a little, but for some users, the devices—and there are numerous imitators—can be part of a routine of staying healthy.

SCIENCE FACTS

- Audex Motorola Cargo Jacket was the first electronic jacket launched in the world.

- Kuchofuku Air-Conditioned Cooling Work Shirt has two built-in fans to keep the body cool.

- Not only do Fitbits help track fitness use, many smartphones have apps that do similar work.

Smart Home Devices

One of the biggest parts of the electronics industry has been called the "Internet of things." With the growth of wireless connectivity, more and more devices are being made to interact with the Internet. Perhaps the biggest area of growth is in the home, where device makers are looking for more and more ways to help people live a fully connected life.

Control Pods

Google, Apple, and Amazon created devices that use voice commands from the user to handle many tasks, most of them connected to the Internet. The Google Home pod (above) can be tasked with monitoring home systems, ordering takeout food, choosing music to play, and making phone calls. Similar apps for smartphones (left) let people control home systems such as heating and air conditioning.

SCIENCE FACTS

- Along with Google Home, Apple's HomePod and Amazon's Alexa are popular digital home controllers.

- One study predicts 36 million "connected" homes in the United States by 2020.

Siri and Alexa at work

Appliances That "Think"

The kitchen has become packed with appliances that use the power of the Internet to (it is hoped) improve people's lives. One of the most connected devices has become the refrigerator. A panel in the door of many models acts as a touchscreen/tablet. Users can play videos, look up recipes, order food supplies, or leave messages. Some models also have a camera that can show what is in the fridge or sensors that relay to the user what food has been used up and should be replaced.

Next-Gen TV

L ED TV (Light-emitting diode television) is an upgraded display technology of the LCD TV (liquid crystal display television). Its major features are its lightweight construction and portability. The LED TV technology is considered environment friendly. It functions more efficiently and has eco-friendly screens. LEDs are power efficient and have a longer lifespan.

Perfect Picture Quality

The picture quality of an LED TV is much better than that of an LCD TV. Since LED TV uses LEDs (light-emitting diodes), the brightness is significantly improved over that of an LCD TV, which uses built-in fluorescent tubes. The darkness and brightness of each can be adjusted in many ways. The LED TV provides its viewers more realistic, **vibrant**, and colorful images.

SCIENCE FACTS

- An LED television life span has been predicted to be one million hours.
- The first LED TV screen was developed in 1977.

Energy efficiency

LED televisions are energy efficient. They consume much less energy as compared to LCD televisions. They **emit** less heat and require much less power to produce an image, significantly lowering the amount of energy used. Its power consumption is almost 50 percent less than that of an LCD television. The image below shows the image quality improvement as the pixels per inch grows.

480p
720p
1080i/p
2160p (4K)

GPS

A GPS (global positioning system) navigation device is used for determining one's current location on earth. A location can be determined because of the GPS satellite signals received by these devices. They are mostly used in military, **aviation**, marine, and consumer product applications. They consist of maps displayed in human readable format via text or in a graphical format.

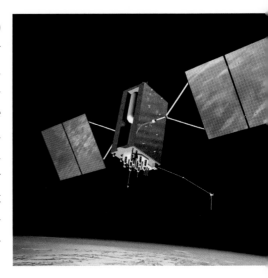

Vehicle navigator

A GPS can be used as a **vehicle navigator**. It can be used for providing information on traffic conditions or nearby amenities like restaurants and fuel stations. For example, if a person is in a new place, then he does not have to worry about getting lost and asking for the directions if his car has a GPS tracking device. They help in determining exact location, calculate distance, and verify the direction that one needs to take in order to arrive at a particular place. GPS tracking devices are now mostly used in boats and ships, in order to make it easy to find directions while sailing.

Application

The most important benefit of a GPS tracking device is just what it implies: "to track" or determine the location of an object or a body. They provide a multitude of applications, benefits, and uses to suit most of the individual needs of its users. They can be used as a tracking device for finding a stolen vehicle, a lost pen, or monitoring the location of endangered species, etc.

Drone Cameras

Look, up in the sky! It's a bird … it's a plane … it's a drone? From a small hobby to a growing industry, small, unmanned flying devices called drones are found everywhere now. When equipped with cameras, they have changed how we view our world.

A Long History

Unmanned air vehicles (UAVs) have been around for more than 100 years. The military uses large-scale UAVs in **surveillance** and in attack. The development of smaller, consumer-friendly UAVs has opened up the world of flight for millions of people.

SCIENCE FACTS

- The US military uses drones to drop bombs, record pictures and video, and to prevent pilot deaths.

- Drones are also used in scientific research, providing views of wildlife and landscape not possible with more standard aircraft.

Eyes in the Sky

The advent of small, portable still and video cameras—such as those made by GoPro—has given drone pilots a new tool. The pilots can control the cameras from the ground to create amazing new films. The cameras are also used to help farmers examine their fields, real estate agents show off property, and search-and-rescue experts save lives.

Connection

The key to the boom in drone usage was getting the size and cost down. Control panels are relatively easy for beginners to learn to use. Some drones can be controlled by touch-screen smartphones as well.

The Future

Technology has advanced rapidly over the past decade and has changed the way people live. Communicating with someone who is sitting across the world is now a simple key press away. What type of technology will govern the lives of people in a hundred years? Maybe cars will not require drivers anymore, and will be automated and respond to voice commands; or, perhaps they will fly in the sky, eliminating traffic jams. We do not know what type of technology we will have in a hundred years, but by the year 2050, we might experience some brilliant advancements in technology.

Nanofibers

Nanotechnology uses microscopic and molecular-sized material to create new and amazing products. By embedding nanotech that can connect to wireless transmission, new types of fabric and building materials might be made that are cheaper and safer. Solar cells are already being made with nanotechnology, possibly helping expand this energy-saving tech.

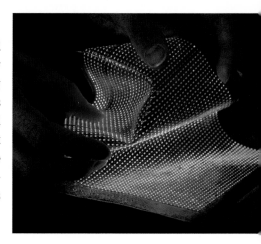

Augmented Reality

Virtual reality, as we've seen, is a fully-immersive tech that puts a person wholly in another world. **Augmented** reality (AR) is sort of the reverse: It puts a virtual world over the real world. Smartphones with AR can "read" the world around them and provide information as an overlay to what the phone camera sees. The popular Pokémon Go game used AR. Travel apps can use it to help people find places to go.

SCIENCE FACTS

- A survey in 2018 found that more than 64 percent of Americans are "very concerned" about driverless cars. Where to do you stand: For or against?

- Interested in a 360-degree VR platform? An early model is selling for nearly $16,000!

Self-Driving Vehicles

Cars without drivers are clearly seen as a wave of the future. Numerous companies are well along in their experiments with the machines. Huge tractor-trailers have been guided across the country without drivers, and cities have let some driverless cars on their streets. The safety factor is still a hurdle and some accidents have happened, but get ready, these cars are coming to your town soon.

VR Treadmill

Wearable VR goggles are one thing. Why not a full-body experience? Researchers are looking for a way to let a person affect their VR environment with their whole body by walking on a treadmill connected to the device. The machine shown is helping an injured soldier undergo rehab by moving through a virtual space on a treadmill. Other ideas include similar machines that let users move through 360 degrees.

TEXT-DEPENDENT QUESTIONS

1. What material was the first computer mouse made from?

2. In what year was the first successful mobile phone made?

3. How does Bluetooth connect devices?

4. Name a device that can use an SD card.

5. Name one of the "connected home" devices mentioned in the text.

6. What does LCD TV stand for?

7. What is another name for drone aircraft?

8. What game does the text mention was played using augmented reality?

RESEARCH PROJECTS

1. Design time: After reading about all the possibilities described in the book—and perhaps after looking into other emerging tech—design a device that you think the world needs. Is it for the home? For school? For business? How does it work? What does it look like? Make some drawings and try to "sell" your friends on it.

2. Imagine you are the pilot of a small drone equipped with a video camera. What do you think you would like to capture images of? Where would you fly it? Before you take off, research local rules about small drones; many cities and towns restrict the use of the devices. Why do you think they do that?

3. Take a poll of your classmates, family, and friends. Ask them what they feel about driverless cars. Then gather their answers into a Pro and Con chart. What did you find out? How did answers differ among your friends? Did your parents or older adults have a different point of view?

Books

Hansen, Dustin. *Game On! Video Game History from Pong and Pac-Man to Mario, Minecraft, and More*. New York: Feiwel and Friends, 2016.

Kallen, Stuart. *What Is the Future of Drones?* San Diego: Reference Point Press, 2016.

Torok, Simon, and Paul Halper. *Imagining the Future: Invisibility, Immortality, and 40 Other Incredible Ideas*. Clayton, Australia: CSIRO Publishing, 2016.

On the Internet

WIRED Magazine
https://www.wired.com/
A great resource for news and information on computers and electronic devices of all kinds.

The smartphone story
https://www.thoughtco.com/history-of-smart-phones-4096585
They're glued to your hands; do you know how they came to be?

Driverless cars?
https://www.weforum.org/agenda/2018/04/driverless-cars-are-forcing-cities-to-become-smart
The World Economic Forum examines the issue.

SERIES GLOSSARY OF KEY TERMS

alloy a substance made up of a mixture of metals

capacitor a device that stores electricity

emission the act of sending out gases or heat into the atmosphere

digital expressed as series of the digits 0 and 1

friction a force that is produced when one object rubs against another object

hydraulic using powered created by water or liquid

interactive providing output based on input from the user

interplanetary a space mission that is planned for study of other planets

magnetic field a region around a magnet that has the ability to attract other magnets

microprocessor a very small circuit used in computers that performs all the functions of a central processing unit (CPU)

navigation the science of directing the course of a vehicle

programmable able to be given instructions to do a task

protocol a set of rules that is used by computers to communicate with each other across a network

renewable something that can replace itself by a natural process

rechargeable something that can be charged again and again

sensor a mechanical device that is sensitive to some signal and helps in responding to it

voltage difference in electric tension between two points

synchronize to operate two or more devices at the same time

viable capable of working successfully

INDEX

Photo Credits